HOW & WHY

INTERFERENCE OF

LIGHT

TAKES PLACE?

BY

NARENDRA SWARUP AGARWAL

Copyright © 2021 Narendra Swarup Agarwal

All rights reserved.

ISBN-13: 9798704160786

DEDICATION

Dedicated to my parents who taught me to observe and analyse nature.

PREFACE

The Interference of Light is a well-known phenomenon and is one of the most important phenomena in Physics.

There are three types of Interference phenomena:

- Constructive Interference
- Destructive Interference
- Intermediate Interference

However, the science of these phenomena: HOW these phenomena take place and WHY is not known.

This book explains the hidden science of all the three types of interference phenomena.

TABLE OF CONTENTS

Introduction	1
New Quantum Theory	3
Young's Double Slit Experiment	7
Interference of Light	13
Constructive Interference	17
Destructive Interference	21
Intermediate Interference	25
Conclusion	29

CHAPTER 1

INTRODUCTION

The scientists worked for more than 400 years to understand and analyse the nature of light.

In 1637, Descartes developed the '**Corpuscular Theory of Light**' which was elaborated by Isaac Newton in 1672. According to this theory, the light is made up of small discrete particles called '**Corpuscles**' (small particles) which travel in straight line with a finite velocity.

Christiaan Huygens proposed a mathematical '**Wave Theory of Light**' in 1678 and published in his Treatise of Light in 1690. Huygens proposed that light was emitted in all directions as a series of '**Waves**' in a medium called the Luminiferous Ether vibrating up and down.

Thomas Young in 1801 proved the **Wave Theory of Light** by the Double Slit Experiment. The actual distribution of brightness can be explained by the alternately additive and subtractive interference of waves. But the light is absorbed as the particles on the screen.

The double slit experiment proved the dual nature of the light known as the Wave-Particle Duality. The light exhibits the behaviour of both the particle as well as wave. Unable to explain the Dual Nature of Light, the scientists in the 20th century accepted Wave-Particle Duality as fundamental nature of light.

The analysis of Double Slit Experiment by Thomas Young explains:

How the three different types of interference phenomena take place?

Why the three different types of interference phenomena take place?

Interference Phenomenon explains the nature of the photons, which has remained a mystery so far.

CHAPTER 2

NEW QUANTUM THEORY

The photons or the quantum particles display mysterious Quantum Phenomena such as Interference, Wave-Particle Duality and Quantum Spin Hall Effect etc. Even though several Quantum Phenomena are known since long, the Quantum Theory is unable to explain any of the Quantum Phenomena. The **New Quantum Theory** developed in the year 2012 in India explains all the mysterious Quantum Phenomena.

The NEW QUANTUM THEORY states:

"A photon/quantum particle has a nucleus of the mass and the charge located off-centre. The nucleus has nearly all the mass and charge of the photon."

A small nucleus of the mass and the charge, located off-centre in the spinning photon/quantum particle, bestows unique characteristics to display strange phenomena.

This constitutional essence of a photon/quantum particle explains not only the Wave-Particle Duality but all the Quantum Phenomena and completes the Quantum Theory.

Figure 1 shows a Photon with a Nucleus of mass and charge which is not in the centre of the Photon.

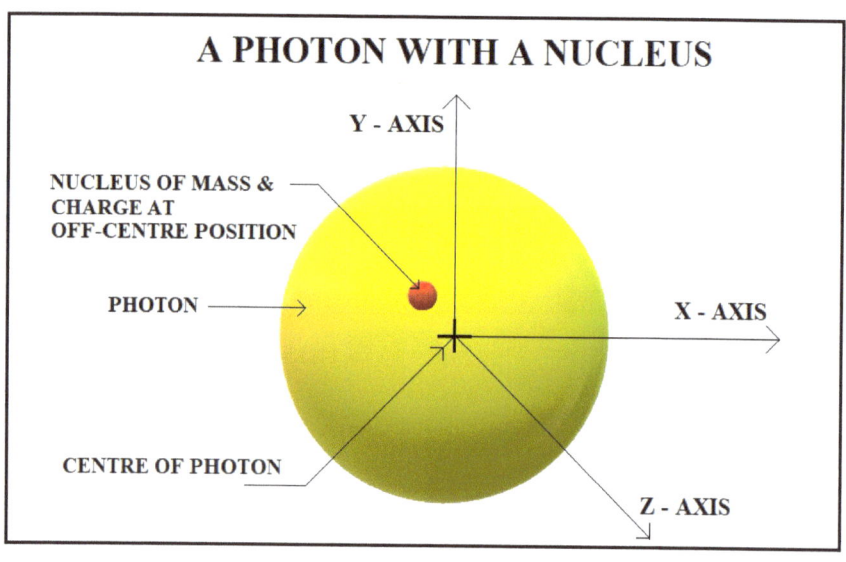

Figure 1: The big yellow sphere shows a photon, and the small red sphere shows a nucleus of mass and charge in the photon. The nucleus is not in the centre of photon but located off-centre in the photon.

How and Why Interference of Light takes place?

As the photon spins, the nucleus of mass and charge of the photon rotates around the centre of the photon. The velocities of the nucleus at θ^0 phase angle of the nucleus are calculated below:

- Horizontal velocity in X-direction : $2\pi r f \sin\theta$ meter/sec
- Vertical velocity in Y-direction : $2\pi r f \cos\theta$ meter/sec

The velocities change continuously with the change in phase angle.

The mass (m) in the nucleus of a photon develops the forces as under:

- **Force F_x in X - direction :**

  $2\pi r f m [\sin(\theta + d\theta) - \sin\theta] / dt$ kg meter/sec^2

- **Force F_y in Y-direction :**

  $2\pi r f m [\cos(\theta + d\theta) - \cos\theta] / dt$ kg meter/sec^2

Note: The detailed calculations are available in 'The Science of Electromagnetic Waves' by the author.

Narendra Swarup Agarwal

CHAPTER 3

YOUNG'S DOUBLE SLIT EXPERIMENT

The **Figure 2** shows a basic setup of the double slit experiment. The sunlight first passes through a Red Filter and subsequently through a slit to achieve a coherent state. This coherent light passes through a pair of slits in the close position. The light from the pair of slits interferes and falls on a screen.

A pattern of bright red and dark interference bands becomes visible on the screen due to the Interference phenomenon.

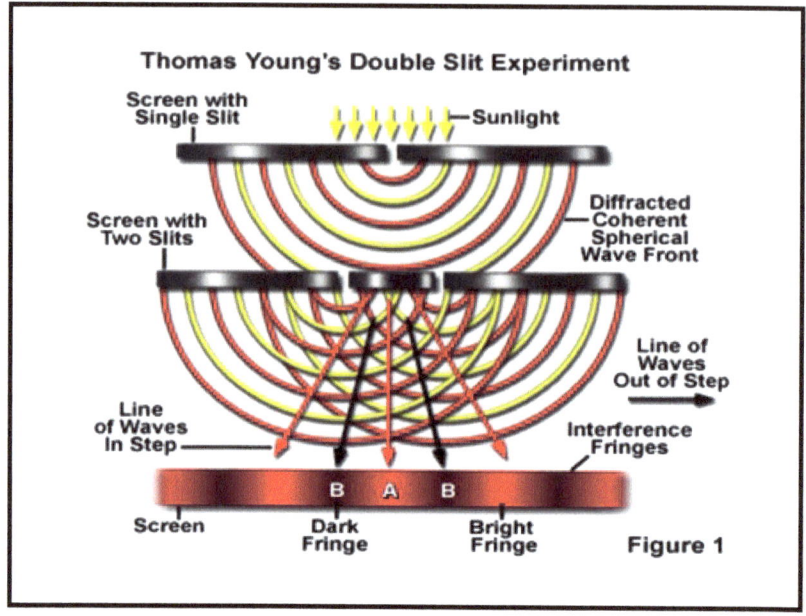

Figure 2: Young's Double Slit experiment

Source:

https://www.olympus-lifescience.com/en/microscope-resource/primer/java/doubleslitwavefronts/

OBSERVATION OF LIGHT PATTERN

The pattern of light on the screen shows:

A bright fringe 'A' in the middle (with reference to the pair of two slits).

Dark fringes 'B' on both sides of the bright fringe 'A'.

There is a regular decrease in the intensity of light from the bright fringe 'A' to the dark fringe 'B' on both the sides.

There is regular increase in the intensity of light from the dark fringe 'B' on both the sides.

ANALYSIS OF EXPERIMENT

The bright fringe 'A' on the screen is at equal distance from both the slits, therefore, the photons of light from the two slits reach there in the same phase and the light intensity is the brightest.

The dark fringe 'B' on the screen is at unequal distance from the two slits so that the phases of the photons from the two slits are in opposite phase. Therefore, the light intensity is the darkest.

The intensity of light from the bright fringe 'A' to the dark fringe 'B' decreases gradually as the difference between the distances from the two slits increase. This indicates the difference in phase angles between the photons from the two slits increase continuously from the bright fringe 'A' to the dark fringe 'B'.

CONCLUSION OF EXPERIMENT

The analysis of the Double Slit Experiment concludes that the photons are not uniform particles but have a nucleus of mass and charge as per the New Quantum Theory. The phase angle of the photon or the angular position of the nucleus in the photon is responsible for the different types of Interferences – Constructive, Destructive and Intermediate Interferences.

If a photon has no nucleus of mass and charge located in off-centre position, the photons can never display different types of interferences.

Narendra Swarup Agarwal

CHAPTER 4

INTERFERENCE OF LIGHT

'Interference' is the one of the most important phenomena of the Quantum Physics.

- The photons of same frequency in the same phase display **Constructive Interference**.
- The photons of the same frequency in the opposite phase display **Destructive Interference**.
- The photons of the same frequency in other different phases display **Intermediate Interference**.

The type of Interference phenomena depends on the 'PHASE OR PHASE ANGLE' of the interfering photons.

As a photon spins, its nucleus of the mass and the charge rotates around the centre of the photon and rotates 360^0 angle around the centre of the photon in one spin of the photon.

PHASE ANGLE:

The angle of 'Nucleus of mass and charge' from the Axis/Direction of travel of the photon is known as **'Phase Angle of the photon or the nucleus'**. The term phase angle may be well known to most of the readers, for the others it is explained below:

A new wave cycle of the photon starts from X – Axis (0, 0) coordinates when the nucleus of mass and charge (red sphere) is at 0^0 phase angle from the X – Axis as shown in **Figure 3**.

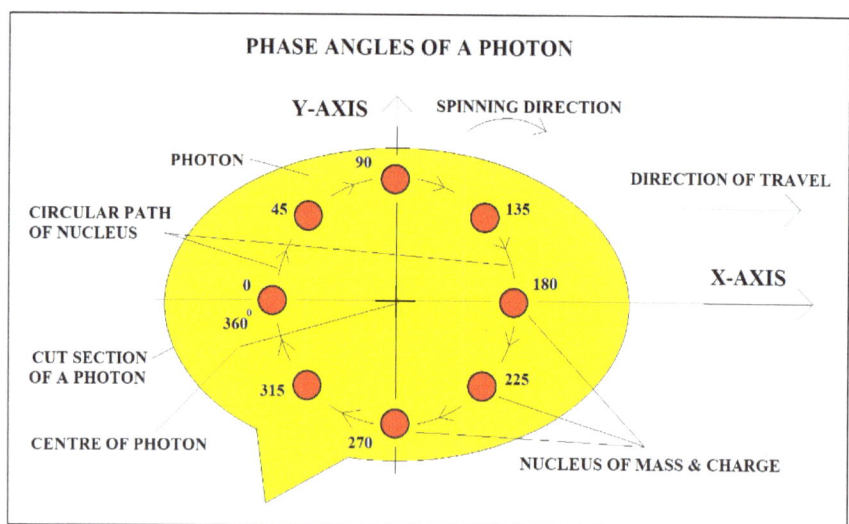

Figure 3: A Cut Section of the central portion of a linearly polarised photon as yellow circle is shown in the Figure.

The photon travels in X-Axis. The red circles show the different phase angles of the nucleus (or the photon). The phase angles of the nucleus from 0^0 to 360^0 are shown with 45^0 increments.

When the nucleus is on the X-Axis on the left-hand side, the phase angle is 0^0. As the photon spins in the clockwise direction, the nucleus rotates, and the phase angle increases continuously and reaches to the maximum value 360^0.

If the two photons of the same frequency and in the same direction superpose, the two photons can join only in side-by-side position. Both the superposed photons keep on spinning in their direction of spin and move together. The mass in each photon being close enough, the strong force is active to keep the superposed photons together.

Figure 4 shows the two spinning photons joined side by side which is the necessary condition for the Interference phenomenon.

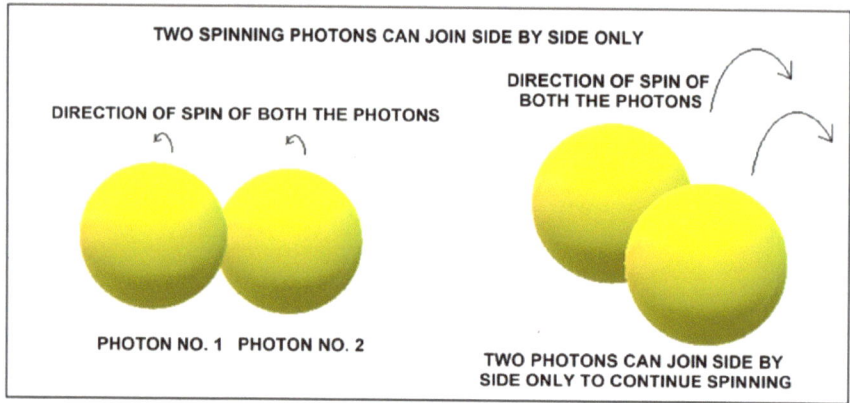

Figure 4: The Figure shows two sets of photons in side-by-side position. Two spinning photons of the same frequency can join only in side-by-side position to Interfere.

CHAPTER 5

CONSTRUCTIVE INTERFERENCE

In the Constructive Interference of the two photons, the amplitude of the resultant wave doubles. When the two photons of the same frequency and in the same phase superpose, the angular positions of the nuclei of both the photons are also the same. Being in the same phase, the mass, in the nuclei of both the photons, accelerate/decelerate and develop the forces of the same magnitude all the time.

The force F_Y in the vertical direction, developed by the mass of each photon in the same phase angle θ, adds to double as under:

The force developed by the nucleus of Photon No. 1 in Vertical Y-axis:

$$2\pi r f m [\cos(\theta + d\theta) - \cos\theta] / dt$$

The force developed by the nucleus of Photon No. 2 in Vertical Y-axis:

$$2\pi r f m [\cos(\theta + d\theta) - \cos\theta] / dt$$

Total Force developed by the nuclei of Photons 1 & 2 in Vertical Y-axis:

$$4\pi r f m [\cos(\theta + d\theta) - \cos\theta] / dt$$

Figure 5 shows two photons in the same phase exhibiting 'Constructive Interference.'

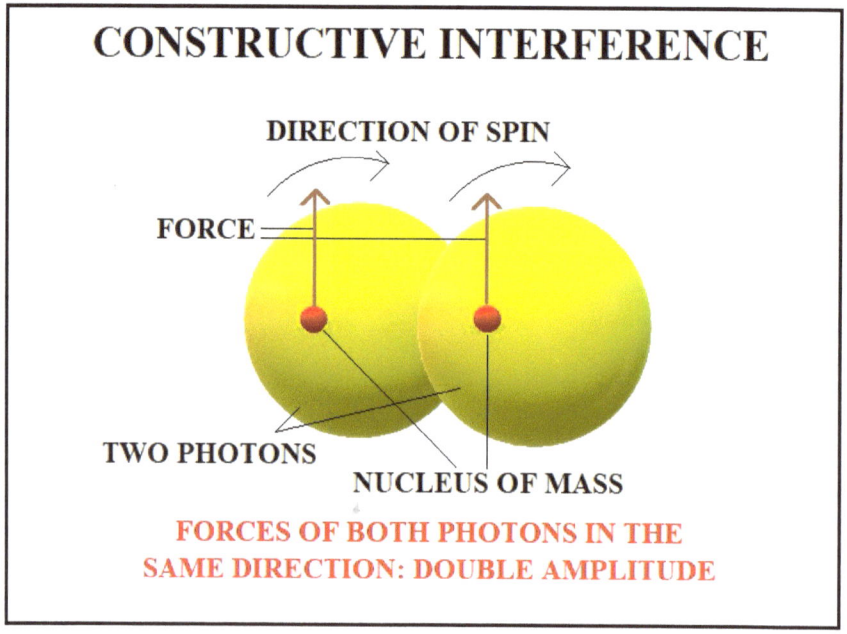

Figure 5: Two interfering photons are in the same phase and joining side by side. The mass in the nuclei of both the photons generate Forces of equal magnitudes in the direction perpendicular to the direction of travel to double the amplitude.

The Constructive Interference proves the presence of the mass in the photons.

The Constructive Interference phenomena is possible if the photons have uniformly distributed mass or the mass in the centre of the photons.

If a photon has zero mass, it cannot develop any force to double the amplitude. Only the presence of a mass in each photon generates the forces of equal magnitude in both the photons to double the amplitude.

THE PHENOMENON OF 'CONSTRUCTIVE INTERFERENCE' WITH THE DOUBLE AMPLITUDE OF THE RESULTANT WAVE DISCOVERS AND PROVES:

"A PHOTON HAS MASS."

Narendra Swarup Agarwal

CHAPTER 6

DESTRUCTIVE INTERFERENCE

In the Destructive Interference, the two photons superpose with a phase difference of π and form the resultant wave with the zero amplitude. The Figure 14 shows the two photons of the same frequency with a phase difference of π in the superposed position developing the forces of equal magnitudes in the opposite directions.

The nuclei of the mass of both the spinning photons have a phase angle difference of π. The directions of the forces developed by the mass in the nucleus of each photon are all the time opposite to each other due to the phase difference of π.

The phase angle of the Photon No. 1 is (θ) and the phase angle of the Photon No. 2 is ($\theta + \pi$) for the Destructive Interference.

The two photons develop forces in the vertical direction as under:

The force developed by the nucleus of Photon No. 1 with phase angle θ in Vertical Y-axis:

$$2\pi r f m [\cos(\theta + d\theta) - \cos\theta] / dt$$

The force developed by the nucleus of Photon No. 2 with phase angle $\theta + d\theta$ in Vertical Y-axis:

$$2\pi r f m [\cos(\theta + \pi + d\theta) - \cos(\theta + \pi)] / dt$$
$$\text{Or:} \quad (-) 2\pi r f m [\cos(\theta + d\theta) - \cos\theta] / dt$$

Total Force developed by the nuclei of Photons 1 & 2 in

Vertical Y-axis: **ZERO**

Both the Vertical Forces are equal but in the opposite directions, therefore, the sum of both the Vertical Forces is all the time Zero. Hence, the net amplitude of both the photons is always Zero in all the phase angles.

Figure 6 shows two photons displaying 'Destructive Interference' phenomena.

How and Why Interference of Light takes place?

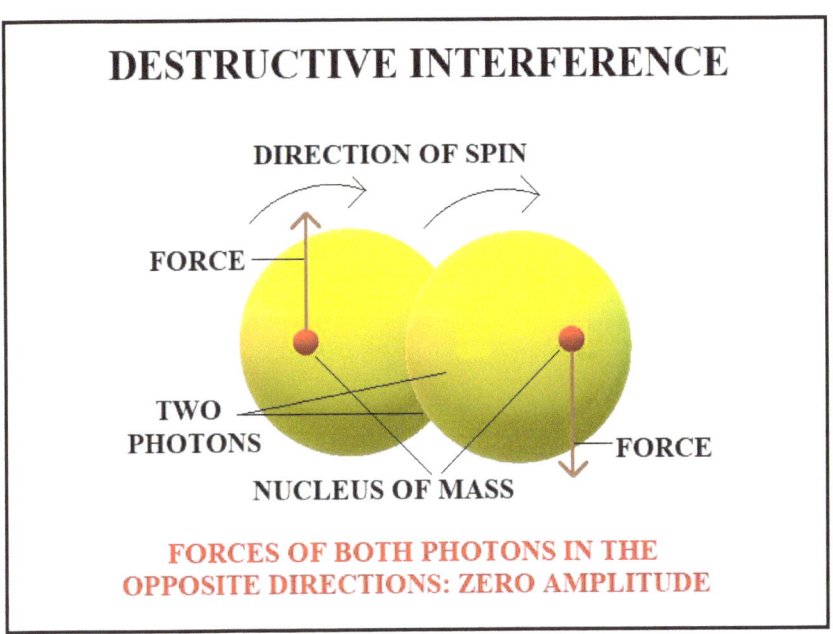

Figure 6: Two photons with phase angles having difference of π. The nuclei of both the photons develop the Forces always in the opposite directions. Both the photons are in side-by-side position.

The Destructive Interference phenomenon is not possible if the mass is uniformly distributed or in the centre of the photon. In such a situation, whatever is the phase difference between the two photons, only the Constructive Interference can take place.

THE DESTRUCTIVE INTERFERENCE IS POSSIBLE ONLY IN A CONDITION:

"THE LOCATION OF THE NUCLEUS OF THE MASS IS IN THE OFF-CENTRE POSITION IN THE PHOTONS."

However, the Destructive Interference phenomenon can also take place even if a photon has more than one nucleus of the mass, located off-centre in the photon.

THE PHENOMENON OF DESTRUCTIVE INTERFERENCE WITH THE ZERO AMPLITUDE OF THE RESULTANT WAVE DISCOVERS AND PROVES:

"A PHOTON HAS THE MASS LOCATED IN AN OFF-CENTRE POSITION IN THE PHOTON."

CHAPTER 7

INTERMEDIATE INTERFERENCE

When the two photons, with a phase difference other than $0°$ or π, superpose the resultant wave with the amplitude > 0 but $<$ double the amplitude forms due to the phenomenon of the Intermediate Interference.

If the two photons of the same frequency with phase angles θ_1 & θ_2 interfere, the mass in the nucleus of each photon develops the force in the vertical direction as under:

The force developed by the nucleus of a Photon no. 1 in the Vertical Y-Axis:

$$2\pi\, r f m\, [\cos(\theta_1 + d\theta_1) - \cos\theta_1] / dt$$

The force developed by the nucleus of a Photon no. 2 in the Vertical Y-Axis:

$$2 \pi r f m [\cos(\theta_2 + d\theta_2) - \cos\theta_2] / dt$$

Total force developed by the nuclei of both the Photons in Vertical Y-axis:

$$2 \pi r f m [\cos(\theta_1 + d\theta_1) - \cos\theta_1] / dt \ +$$

$$2 \pi r f m [\cos(\theta_2 + d\theta_2) - \cos\theta_2] / dt$$

The amplitude of the resultant wave depends on the values of the phase angles θ_1 & θ_2 of the two photons.

Figure 7 shows two photons displaying 'Intermediate Interference' phenomena.

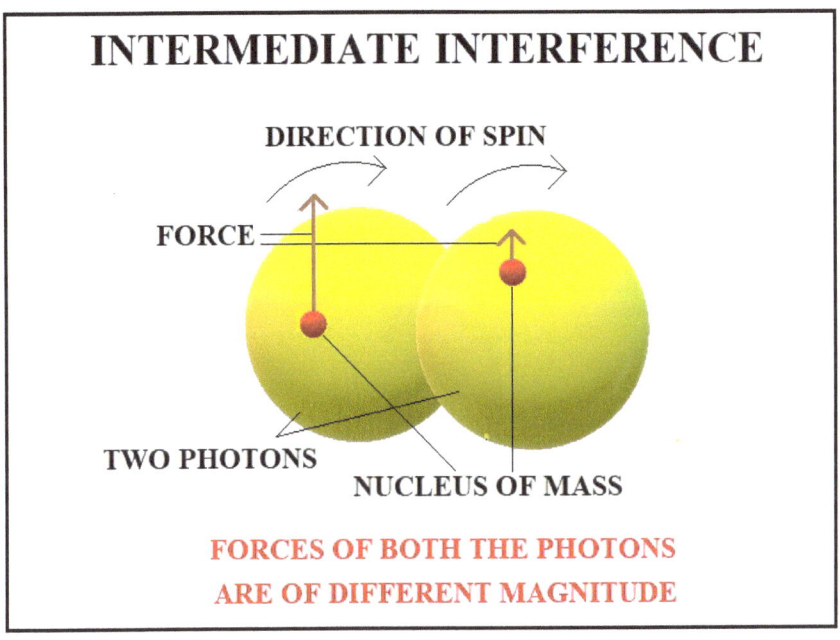

Figure 7: The two photons develop forces of unequal magnitudes in the same or different direction by the nucleus of mass being in the different phase angles (other than the difference of π or 0).

If a photon has more than one nucleus of the mass, the two photons on intermediate interference will form a resultant wave with two or more peaks.

However, the resultant wave has only a single peak as observed in the Double Slit Experiment. The intensity of band changes steadily from bright to dark band and vice versa.

THE SINGLE PEAK IN THE RESULTANT WAVE ON INTERMEDIATE INTERFERENCE DISCOVERS AND PROVES:

"THE PRESENCE OF ONLY ONE NUCLEUS OF THE MASS LOCATED OFF-CENTRE IN A PHOTON."

THE THREE TYPES OF INTERFERENCE PHENOMENA DECISIVELY PROVE THAT A PHOTON HAS ONLY ONE NUCLEUS OF MASS LOCATED AT OFF-CENTRE POSITION IN THE PHOTON.

THE INTERFERENCE OF PHOTONS PROVES THE NEW QUANTUM THEORY.

CHAPTER 8

CONCLUSION

The two photons of the same frequency interfere and display Constructive or Destructive or Intermediate Interference phenomenon depending the phase angles of the two photons.

Constructive Interference: The two photons have the same phase angle, therefore, the nucleus of mass in the photons generates forces of equal magnitude in the same direction.

Destructive Interference: The two photons have the phase angle difference of 180^0, therefore, the nucleus of mass in the photons generates forces of equal magnitude in the opposite directions.

Constructive Interference: The two photons have the phase angle difference except 0 or 180^0, therefore, the nucleus of mass in the photons generates forces of unequal magnitude in the different directions.

www.ingramcontent.com/pod-product-compliance
Lightning Source LLC
Chambersburg PA
CBHW040255220526
45473CB00001B/493